Software Requirements Specifications

A How To Guide for Project Staff

David Tuffley

To my beloved Nation of Four
Concordia Domi – Foris Pax

There is no fair wind for one who knows not whither he is bound.
-- Lucius Annæus Seneca, 3-65 AD

Acknowledgements

I am indebted to the Institute of Electrical and Electronics Engineers on whose work I base this book, specifically IEEE Std 830 Guide to Software Requirements Specifications, and IEEE Std P1233 Guide for Developing System Requirements Specification. Also the *Turrbal* and *Jagera* indigenous peoples, on whose ancestral land I write this book.

Contents

Contents

Contents

A.   Introduction

Developing Software Requirements Specifications: A Guide for Project Staff outlines how project staff can develop:

- Requirements Lists(RL),

- Statement of User Requirements (SUR) and

- Software Requirements Specifications (SRS).

The end product of the requirements capture process is the complete and accurate definition of the functionality of the proposed system. It is a 'top down' process which proceeds from the general to the specific through a series of predefined steps.

This book gives a detailed outline of the requirements capture process. It discusses how to apply the steps contained in the requirements capture standards (IEEE Std 830 Guide to Software Requirements Specifications, IEEE Std P1233 Guide for Developing System Requirements Specification).

A.1. How this book is organised

Part A: Introduction

Part B: Requirements List (RL)

Part C: Statement of User Requirements (SUR)

Part D: Software Requirements Specification (SRS)

Part E: References

A.2. The nature of requirements capture

Developing a comprehensive understanding of user requirements is one of the most problematic aspects of the entire software development process.

High quality software can only be developed based on such an understanding, and this calls for close cooperation and understanding between users and developers, particularly during the requirements gathering and analysis stage. The challenge for any developer is to recognize the interdependency of the stakeholders and to work towards creating the conditions in which the sub-systems work harmoniously with each other.

Software developers by nature have a technological mindset. Years of technical education and on-the-job training which develops their technical skills to a higher standard? The pursuit of technical excellence is a matter of professional pride, as it should be. Conversely though many if not most software users have a *non-technical* or limited technical view of the world. Their interaction with software is a means to an end, not an end in itself.

Bringing users and developers into closer cooperation through improved mutual understanding is in the best traditions of Socio-Technical Design. There should be substantial user involvement in the system design process [1]. In this book, systems are defined broadly. A system can be comprised of networks of users, developers, information technology at hand, and the environments in which the system will be used and supported [1].

Achieving this closer cooperation and mutual understanding requires an understanding of the dynamics of the organisational culture in which systems development is performed.

A.2.1. Historical view of user-developer cultural differences

Organisational culture is a useful perspective within which to explore the nature of the User-Developer gap. Culture is an organization's way of thinking about the world and itself, how to get things done, how to solve problems. Software developers living in a world of technology have their own cultures, their own deeply ingrained ways of doing things. Such a culture might find it difficult to readily understand, much less embrace the seemingly foreign culture of the business user, with their unfamiliar priorities, preoccupations, and ways of doing things. It is a case of 'same planet, different worlds'.

An organisation develops is own unique culture over time, evolving through the stresses and strains of its day-to-day operations. Awareness by developers of the mechanisms of organisational culture may well be limited though. People simply go about their jobs in the way they have always done, and do not think much about it until an external threat to their security and continued existence is perceived. At this point, organisational culture activates in-built defence mechanisms that seek to neutralise the perceived threat.

Software developers apparently possess characteristics that inhibit their working relationship with other members of the organisation. As far back as 1991, Grindley [2] surveyed IS directors and found that 46% reported that the culture gap between IS professionals and business counterparts was their most important challenge in terms of service delivery. 56% believed that the culture gap inhibits their organization's ability to achieve strategic advantage using IS. 56% is a startlingly high figure. And let us remember that CIO's or IS Directors are in a unique position to evaluate the effectiveness or otherwise of their department.

Grindley explains that the culture gap is manifested by users and developers having differing approaches to motivation, goals, language, and problem-solving. These differences brought about not only difficulties in communication, which is an overt manifestation of the gap, but also reveals that the mind-set is likely to be different. Mind-set can be said to be a covert manifestation of the cultural gap. Having different notions of goal-setting and problem-solving are indicative of these differences. Grindley's findings are consistent with even earlier findings of researchers such as Edstrom [3], Gingras and McLean [4], and Zmud and Cox [5] when they reported on the distinctive ways of thinking and acting of IS professionals.

Wang [6] defines the culture gap as 'a conflict, pervasive yet unnatural, that has mis-aligned the objectives of executive managers and technologists and that impairs or prevents organizations from obtaining a cost-effective return from their investment in information technology'. This statement highlights the nature of the gap in terms of a misalignment of objectives. This is in agreement with Grindley's earlier study that discussed the problem in terms of different approaches to goal setting, problem solving and language. This misalignment of objectives causes impairment of an organization's achievement of cost-effective systems development because the two categories of stakeholder are pulling in different directions.

Good communication between IS professionals, IS staff and IS users is critical to the successful completion of an IS development project [7]. The ability to interact with all potential stakeholders in an organisation, to clearly document requirements, and to effectively express ideas has long been recognized by researchers and practitioners as critical success factors.

So we are left with the sometimes unbridgeable gap between developers and users that makes successful completion of projects a near impossibility. How best to understand the gap?

A.2.2. Categorizing the gap

The gap has been categorized in no less than nine ways [8]:

The Perspective Gap when the point-of-view of one stakeholder group is incomplete or ill-conceived. Developers may lose sight of the necessity for Systems to provide value to the business by meeting evolving business goals and that the IT department is not the centre of the universe. Users sometimes lose sight of technology as being a tool, and not an end in itself.

Ownership Gap: where developers feels a sense of proprietary ownership over the infrastructure, while users feel ownership over the business processes, leading to the demarcation disputes and territorial conflict that strain the relationship and create misunderstandings and misconceptions. Users can get the impression that developers are technical elitists, and developers come to see users as reactionary detractors.

Cultural Gap: when the stakeholder groups display different traits, values, working behaviours, and/or priorities due to each group attracting certain kinds of person, or acculturate members in the group. Developers tend to be more introverted and analytical, using rational persuasion to influence others. Business users are usually more extroverted, intuitive and use more sophisticated influence strategies. Both users and developers tend to adopt the culture of their respective professions.

Foresight Gap: where one stakeholder group has greater insight into how the future might unfold, but is unable to communicate that vision convincingly to the other stakeholder group. Developers may be well placed to foresee that a user proposed solution cannot work from a technical point-of-view. Alternately users may be better at determining that a developer proposed solution will not be acceptable to them, or will have a negative impact on some aspect of their operations.

Communication Gap: where one stakeholder group simply fails to understand what the other is saying. It is often said by users that Developers have an impenetrable jargon, yet it is also observed that the users may well have their own well-developed jargon. Developers find it difficult to translate the user needs of business units into useful productive systems because they do not understand the business processes and underlying rationale for them..

Expectation Gap: where users have unrealistic expectations about what developers are feasibly able to do. Users have come to expect more from systems because they have generally become more computer literate, or because they have become accustomed to the sometimes heroic efforts of developers to deliver the goods. At the same time, developers are sometimes known to make overblown claims as to what they can deliver, expecting all users to be technologically naïve.

Credibility Gap: where the past performance of developers has been substandard. This is often attributable to failed development projects, or poor customer service such as a not very helpful helpdesk. From the developer's perspective, they may have found users to be overly demanding and/or resistant to change.

Appreciation Gap: where one stakeholder group implicitly feels unappreciated by the other. Developers may form a view that their hard work, long hours and contributions to the organization go unappreciated except when something goes wrong. There is some suggestion that developers, in some cases, wish to be more involved in business planning, but are not invited to do so.

Relationship Gap: where the stakeholder groups do not interact with sufficient frequency to be able to form a viable, constructive relationship as the basis for ongoing work. This might be reinforced by entrenched preconceptions about the other group.

Communication is implicated to some extent in all nine categories above. The Communication Gap (No. 5) is explicitly about communication, while the Perspective Gap, the Ownership Gap and the Foresight Gap and others might be describing inner states of mind; they are externalised to other parties by the communication process.

An examination of these categories highlights the importance of developing effective communications strategies, an area that will be examined the next section.

User-developer communication strategies can be seen as a function of organisational culture. It might therefore useful to examine an appropriate model for understanding organisational culture such that effective communication strategies might be developed.

A.2.3. Facilitator to bridge user-developer gap

Users with their general, non-technical perspective and developers with their almost exclusively technical perspective do not inhabit the same worlds as each other, despite their being in the same physical location. They do not share a common world view and consequently do not possess a common language with which to describe their worlds to each other. A facilitator of communication to act as a bridge, an interpreter between the stakeholder groups is needed to bring about this common understanding and effectively close the gap.

Where available, business analysts traditionally perform the role of facilitator of communication between the business interest groups and developers. But business analysts are not present on every project. In the absence of a business analyst an analyst-programmer or technical writer might effectively be substituted in the role.

A.3. Outline of the requirements capture process

This section is a conceptual overview of the requirements capture process. It is included here for reference. Detail of the actual steps is included in subsequent sections.

A.3.1. Overview

Requirements specification capture documents are developed by the following process.

- Major functional aspects of the proposed system are captured. This is done in consultation with the user - resulting in the requirements list (RL).

- System environment and business processes are documented.

- Functions/transactions are identified in the statement of user requirements (SUR).

- Acceptance criteria are identified in the statement of user requirements (SUR).

- Complete specification of all system requirements are prepared and outlined in the software requirements specification (SRS).

- Screen and report contents are identified in the software requirements specification (SRS).

- User acceptance of the software requirements specification (SRS).

- The software requirements specification is issued.

A.3.2. Organisation & staffing

The customer shall provide at least one competent full-time user representative to assist the team who are developing the requirements list, statement of user requirements and software requirements specification.

The user representative shall coordinate the activities of the actual users who are to have input into the requirements capture process. He/she is responsible for ensuring that the requirements documented in the list are complete and correct.

Where a project is quite large, it may be more practical to break the project into major functional areas and for smaller teams comprising both the developer's staff and a user representative to work on each.

The customer shall also provide other user staff that has input into the Requirements Specification documents on an as-required basis.

A.3.3. Review requirements specification documents

The requirements specification documents shall be reviewed by the user representative and the development team representative.

Internal reviews must be approved before the final customer review.

A.3.4. Accept requirements specification documents

The approved requirements specification documents shall be signed by all parties concerned in the place provided on the title page.

The customer shall then review the requirements specification documents and will do one of the following:

- Indicate their acceptance by signing in the space provided on the cover page.

- Raise the problem issue with the developer to identify defects or deficiencies in the Requirements Specification documents.

For large projects, as mentioned above in 'organisation & staffing', projects can be broken down into major functional areas and for smaller teams to work on each area. Where this is the case, and the resulting specification documents are relatively large, it may be more practical for the user representative from each team to be responsible for the acceptance of the documentation generated by that team.

The user representative is walked through the requirements specification - is given the opportunity to consider carefully before signing it.

A.3.5. Issue requirements specification documents

When the customer has accepted the requirements specification documents the project manager shall authorise the issue of the requirements specification documents. The authorised documents become the baseline for design and acceptance testing.

This baseline controls any subsequent amendments and enhancements in the remaining stages of the development cycle.

In large systems the baseline has the potential to become very complicated. While it is always important to use the correct change control, it becomes vitally important when the baseline is very complicated.

A.3.6. Changing the requirements specification documents

If the customer wishes to amend the requirements specification documents after acceptance, they must raise a 'change request'. The amended documents shall be reviewed and accepted as a new release of the requirements specification documents.

Any changes must be documented and controlled.

A.3.7. Traceability matrices

The above process is monitored by a series of traceability matrices. The matrices are as follows.

- **Requirements capture** - the user requirements specified in the requirements list are mapped to the items of the software requirements specification. Traceability through design and testing is documented in separate traceability documents.

- **Design** - the items of the software requirements specification are mapped to the design documentation.

- **Test** - the items of the software requirement specification are mapped to the test procedures and cases.

B.Requirements List (RL)

The *Requirements List* is the means by which the customer defines for the development group/organisation what they want the new/revised system to do. It is a *general* list of the capabilities, features and characteristics of the system in 'business' terms.

The information provided by the customer is the starting point for identifying every requirement. The aim is to solicit **all** requirements and to ensure that each requirement is stated once and that none are missed.

B.1. Content

It is important that the Customer and any development personnel set-aside any preconceived ideas regarding solutions or "*how*" the new system will be developed. Instead, the emphasis should be on the non-technical, business-oriented, aspects of "*what*" is required for the customer's business.

The table below is a practical way to capture and arrange the essential information for the Requirements List:

RL Ref	Brief Description	Source	Priority	Critic- ality	X-ref DFDs, Supporting Narrative, etc.	Linked to (Ref)

B.1.1. Reference Number

The Reference No. can be any unique combination of letters and numbers which are meaningful to the Customer. For example, "FR" - Functional Requirement followed by an incremental number or "TR" - Technical Requirement followed by an incremental number.

B.1.2. Brief Description

A single sentence or a few words to indicate the essence of the requirement. Further narrative can be included in supporting documentation.

B.1.2.1. Source

The name of the contact person or group for further clarification of the requirement, if necessary.

B.1.2.2. Priority

Use **H**(igh), **M**(edium) or **L**(ow) to indicate those requirements that are to be part of the current release (high) and those that can be part of subsequent releases of the product (medium or low) if the need arises.

B.1.2.3. Criticality

Use **M**(andatory), **D**(esirable), or **O**(ptional) to indicate the relative importance of the requirement to the business. This scheme distinguishes classes of requirements as follows.

- **Mandatory** implies that the software will not be acceptable unless these requirements are provided in an agreed manner.

- **Desirable** implies that these requirements would enhance the software, but would not make it unacceptable if they were absent.

- **Optional** implies a class of functions that may or may not be worthwhile, which gives the supplier the opportunity to suggest something which exceeds the Requirements Specification.

B.1.2.4. X-ref

Include a cross reference to data flow diagrams, supporting narrative, etc. Any separate supporting information should be identified by a unique reference Id.

B.1.2.5. Link

Indicate any other requirement reference that is linked in some way to this requirement. That is, indicate if this requirement is a primary or supporting requirement and to which requirement(s) it is related.

B.1.3. Procedure

The requirements list is a mandatory step in the requirements capture process.

The following procedure applies.

B.1.3.1. Review

The development group and the customer shall *review* the requirements list to determine its degree of completeness. The review can be conducted either as an informal meeting or a formal. The choice is by mutual consent.

B.1.3.2. Approval

After a successful review, the requirements list shall be inspected.

When ready, the approval sheet of the requirements list document is signed by the responsible parties.

B.1.3.3. Change control

It is essential that any changes to the requirements list be documented and controlled.

C. Statement of User Requirements (SUR)

C.1.1. Introduction

A *Statement of User Requirements* document is a high-level narrative of requirements to be addressed by the new system, as described in the Requirements List (see previous chapter). It incorporates knowledge of the current organisation and system model. While it is a step closer to defining requirements in 'developers' terms, it is still a document that customers are able to understand.

The statement of user requirements document is very useful when conducting a feasibility study. It enables the customer to confirm that the requirements outlined in the requirements list have been correctly interpreted by the development group.

C.1.2. Content

This section lists the mandatory headings for the Statement of User Requirements and provides implementation guidelines. The table of contents should be as follows:

Contents
1. Project Background
2. System & business scope
3. System overview
4. Interfaces
5. Outputs
6. Inputs
7. Processing
8. Data storage
I Appendixes   (Optional)

The following sections provide detail of each of the headings making up the table of contents.

C.1.2.1. Project background (section 1)

A brief narrative outlining the historical context of the project.

- What circumstances combined to give rise to the project.

- Reason(s) for the initial project request.

This section should leave the reader in no doubt as to why the project is being undertaken.

C.1.2.2. System & business scope (section 2)

From a business point of view, describe the following.

- The boundaries of the system.

- What is covered.

- What is not covered.

- What business advantage is likely to be derived.

C.1.2.3. System overview (section 3)

A high-level description of the proposed system, indicate the following.

- Where it fits with pre-existing systems - use diagrams where applicable.

- It's relative importance to the customer's business.

- Any system constraints.

- Any significant risks.

- The 'window of opportunity', if applicable.

- Security considerations.

- Data protection matters.

- System accessibility factors.

- Operator constraints.

- Existing controls and new controls to be included.

C.1.2.4. Interfaces (section 4)

Give detail of other systems that would interact with the proposed system.

Where applicable, provide information on the following.

- Human interface(s).

- Hardware interface(s).

- Software interface(s).

- Communications interface(s).

C.1.2.5. Outputs (section 5)

High level description of the system outputs, including the following.

- Business events (i.e. end-of-month activities).

- Reports.

- Data for export to other system(s).

Specify in concrete terms the time constraints and volumes for each output. Use specific measurements or ranges that can be tested and verified.

C.1.2.6. Inputs (section 6)

In the same manner as Outputs, give a high level description of the system inputs, include the following.

- Business events (i.e. end-of-month activities)

- Input screens

- Data imported from other system(s)

Again, specify in concrete terms the time constraints and volumes for each output. Use specific measurements or ranges that can be tested and verified.

C.1.2.7. *Processing (section 7)*

From a business point of view, describe the following.

- Special processing to be applied.
- Formulae.
- Calculations.
- Any applicable standards.
- Performance issues.

C.1.2.8. *Data storage (section 8)*

List existing main data items to be stored and/or databases.

- Name and description.
- Physical location.
- Volume: current and growth predictions.

List new files and/or databases.

- Name and description.
- Size and growth rate.

C.1.3. Procedure

The following procedure applies.

C.1.3.1. Review

The development group and the customer review the statement of user requirements to confirm that it is accurate and complete - a true reflection of the customer's requirements. The review can be conducted either as an informal meeting or as a formal walkthrough.

Since the document forms the basis for the solution model developed in the next phase, it is vitally important that the customer be closely involved in the review.

In the normal course of events, the statement of user requirements shall be the subject of a number of walkthroughs, depending on the number of iterations required to gain an agreement with the customer.

C.1.3.2. Approval

After the successful conclusion of the review process, the statement of user requirements shall be inspected.

The following criteria are checked.

- All requirements specified in the requirements list have been classified and documented in the requirements traceability matrix and in the statement of user requirements.

- Enough supporting information is provided regarding the scope and measurement of requirements.

- All requirements with high priority or high criticality have been noted in the statement of user requirements.

When ready, the approval sheet of the statement of user requirements is signed by the responsible parties.

C.1.3.3. Change control

It is essential that any changes to the statement of user requirements be documented and controlled. Careful attention to change control avoids the possibility of misunderstandings later.

C.2. Interface control document (ICD)

The interface control document is an optional step in the requirements capture process.

Where used, the following procedure applies.

C.2.1.1. Review

The development group and the customer reviews the interface control document to confirm that it is accurate and complete. The review can be conducted either as an informal meeting or a formal walkthrough.

C.2.1.2. Approval

After the successful conclusion of the review process, the interface control document shall be inspected, where the following criteria are checked.

- **Introduction** - sufficient detail regarding the overview, purpose, scope, definitions, acronyms and references is given.

- **Interface description** - detailed descriptions of the general requirements, interface diagrams, assumptions and dependencies are given.

- **Hardware Interface(s)** - detailed descriptions of the hardware interface(s) is given.

- **Software Interface(s)** - detailed descriptions of the software interface(s) is given.

- **User Interface(s)** - detailed descriptions of the user interface(s) is given.

When ready, the approval sheet of the interface control document is signed by the responsible parties.

C.2.1.3. Change control

It is essential that any changes to the interface control document be documented and controlled.

C.3. Software requirements specification

The software requirements specification is a mandatory step in the requirements capture process.

The following procedure applies.

C.3.1.1. Review

The development group and the customer reviews the software requirements specification to confirm that it is accurate and complete - a true reflection of the customer's requirements.

The review can be conducted either as an informal meeting or a formal walkthrough.

In the normal course of events, the software requirements specification shall be the subject of a number of walkthroughs, depending on the number of iterations required to gain an agreement with the customer.

C.3.1.2. Approval

After the successful conclusion of the review process, the software requirements specification shall be inspected, where the following criteria are checked.

Part C: Statement of User Requirements (SUR)

- **Introduction** - sufficient detail regarding the overview, purpose, scope, definitions, acronyms and references is given.

- **General description** - detailed descriptions of the product functions, user characteristics, general constraints and assumptions/dependencies are given.

- **Specific requirements** - detailed descriptions of the functional requirements, including introduction, input, processing and output is given.

- **External Interface(s)** - detailed descriptions of the external interface requirements is given.

- **Performance requirements** - detailed descriptions of the performance requirements is given.

- **Design constraints** - detailed descriptions of the design constraints is given.

- **Security requirements** - detailed descriptions of the security requirements is given.

- **Maintainability requirements** - detailed descriptions of the maintainability requirements is given.

- **Reliability requirements** - detailed descriptions of the reliability requirements is given.

- **Availability requirements** - detailed descriptions of the reliability requirements is given.

- **Database requirements** - detailed descriptions of the database requirements is given.

- **Documentation requirements** - detailed descriptions of the documentation requirements is given.

- **Safety requirements** - detailed descriptions of the safety requirements is given.

- **Operational requirements** - detailed descriptions of the operational requirements is given.

- **Site adaptation** - detailed descriptions of the site adaptation requirements is given.

When ready, the approval sheet of the software requirements specification is signed by the responsible parties.

C.3.1.3. Change control

It is essential that any changes to the software requirements specification be documented and controlled. A rigorous approach to change control avoids the possibility of misunderstandings later.

C.4. Sign off

When the software requirements specification and interface control document (if done) have been reviewed and approved they are signed off by the developer and customer representatives.

D. Software Requirements Specification (SRS)

The *Software Requirement Specification* is the major deliverable arising from the requirements capture process. For optimal results, the SRS should be based on the preliminary work of *Requirements List* and *Statement of User Requirements* (see previous chapters).

This process consists of four stages.

- **Problem analysis** - where an analyst team interviews the users to establish what problems are to be solved without defining how they will be solved (Requirements List).

- **External product description** - where analysts and users describe the external behaviour of the product without defining how it will work internally. This occurs in either the business or the operations environment.

- **Conceptual design** - where the major functional and/or physical components of the system are identified.

- **Software Requirements Specification** - where external behaviour of the software components of a system are described without regard to how they work internally.

In all but the smallest of systems, requirements engineering is an iterative process involving requirements definition followed by conceptual design (component definition), followed by further requirements definition for each component. The end result is a hierarchy of components with associated requirements specifications. The specifications may be contained in a single document or a set of documents (one for each component). This process simplifies problem definition by progressively partitioning a problem into its component sub problems. Each smaller sub problem can then be analysed in isolation.

D.1. Characteristics of an SRS

The basic feature of an SRS is that it should specify the results to be achieved by the software, not the means by which those results are obtained (i.e. Design). Specify the functionality, performance and design constraints without embedding any design specifications.

This section outlines the desirable properties of an SRS. The absence of any of the properties described below may be viewed as a defect in the context of an SRS document inspection.

D.1.1. Correct

The degree to which a documented statement of requirement adheres to stated facts. For example, if the customer states that the system is required to detect a 20 millisecond pulse and the SRS states 'the data acquisition system shall be capable of detecting a 20 millisecond pulse' - the requirement is incorrect because it embeds a design statement - "the data acquisition system".

Note: The correctness of an SRS can only be improved by a thorough review with the customer.

D.1.2. Unambiguous

Each requirement statement has only *one* semantic interpretation. As a minimum, each characteristic of the final product shall be described in a single unique term.

In practise this means that if ten people were given the same sentence they would interpret it in the same way. If a natural language description introduces , simple requirements modelling tools may be used.

D.1.3. Complete

The degree to which the SRS records *all* of the desired behaviours of the system. More specifically, a SRS is complete if it thoroughly addresses the following four issues:

- **Functions** - everything that the software is supposed to do is included in the SRS. Omissions may be detected by thorough review of the SRS document with the customer.

- **Stimulus/response** - definitions of all responses of the software to all realisable classes of input data under all realisable classes of situations should be included. **Note:** It is important to specify the responses to both valid and invalid inputs. Responses to inputs must also be viewed in the context of various system states.

- **Document housekeeping** - all pages are numbered and identified as per the standard, all figures and tables are numbered named and referenced, all terms are defined, all units of measure are provided and all referenced material and sections are present.

- **TBDs** - no sections are marked 'to be determined (TBD)'. If the insertion of a TBD is unavoidable it should be accompanied by a notation of who is responsible for its resolution and by when.

D.1.4. Verifiable

A requirement is verifiable if there exists some finite cost-effective process with which a person or a machine can check that the actual as-built software product meets the requirement. In practical terms this means that statements of requirement must be specific, unambiguous and quantitative. For example:

- **Nonverifiable** - the system shall respond to user queries in a timely manner.

- **Verifiable** - the system shall respond to a user request for an account balance in five seconds or less under operating conditions.

In general SRS authors should avoid the use of subjective adjectives such as **suitable, appropriate** and **timely**.

D.1.5. Consistent

An SRS is consistent if there is no conflict between subsets of requirements.

Examples of inconsistencies include the following:

- **Conflicting terms** - two terms are used in different contexts to mean the same thing. (i.e. speed and velocity).

- **Conflicting characteristics** - two parts of the SRS demand that the product exhibit contradictory traits.

D.1.6. Modifiable

An SRS is modifiable if its structure and style are such that any necessary changes can be made easily, completely and consistently.

Modifiability involves the following:

- **Organisation** - coherent, logical organisation, including a table of contents and cross references.

- **No redundancy** - the same requirement should not appear in more than one place.

D.1.7. Traceable

An SRS is traceable if the origin of each of its requirements is clear (backward traceability) and if it allows the referencing of each requirement in future documentation (forward traceability).

Both classes of traceability shall be provided as follows:

- **Forward** - this allows the designer to demonstrate that he/she has produced a design that satisfies all the customer's requirements. To this end all statements of requirement shall be named and identified with a paragraph number such that they may be referenced in future requirements and design documents.

- **Backward** - this demonstrates that the SRS author knows why every requirement in the SRS exists. Each statement of requirement shall explicitly reference its source in previous documents. If a document does not exist or is not available the individual who stated the requirement shall be identified.

When a requirement is an apportionment or a derivative of another requirement, both forward and backward traceability should be provided. Examples include the following:

- The allocation of a response time to a database function from the overall user response time requirement.

- The identification of a report format with certain functional and user interface requirements.

- A software product that supports legislative or administrative needs (i.e. passenger safety). In this case, the exact legal requirement should be specified.

Forward traceability is particularly important when the software product enters the operations and maintenance phase. As code and design documents are modified, it is essential to be able to determine the complete set of requirements that may be affected by those modifications.

D.1.8. Usable during the operations & maintenance phase

The SRS must address the needs of the operations and maintenance phase, including the eventual replacement of the software.

Maintenance is often done by someone other than the original developer. Local changes can be implemented by means of well-commented code. For changes with a broader scope, however, the design and requirements documentation is essential.

D.1.9. Standards compliant

The SRS must address each issue specified in this book. In practical terms this means the following:

- All specifications shall have the headings specified below.

- If a paragraph heading is not applicable to the user's needs it is to be included and marked not applicable.

- All applicable paragraphs shall address the checklist of key points covered in each subsection of this standard.

D.2. Level of abstraction

D.2.1. Problem partitioning

A common problem faced by the requirements engineer is establishing the level of detail, and consequently the level of abstraction, to be provided in an SRS. Requirements engineers are often hampered by customers who wish to supply either too little or too much detail. Where too little detail is forthcoming, there are complaints that "we are designing the system for you" while the customer who wants to provide too much detail is in effect wanting to be involved in the design process.

The level of detail to be presented to the designer must be consciously set in the context of each requirements specification exercise. As the level of detail increases the SRS becomes less abstract (i.e. the level of abstraction decreases). This amounts to deciding where the external behaviour of the software will be constrained by the customer and where it will be left to the imagination of the designer.

The level of detail may be formally quantified by viewing the problem as a hierarchy of interrelated sub-problems that may be represented as you might a family tree. The most abstract statement of the problem is represented at Level 1 or root of the tree and progressively decomposed into sub problems.

D.2.2. Issues constraining level of abstraction

The point at which requirements end and design begins is still a hotly debated subject. This is because it will necessarily vary in the context of each requirements definition exercise.

What criteria can be used to determine an acceptable level of abstraction for a particular SRS?

The following factors provide guidelines:

- **Meeting business objectives** - if the omission of certain required behaviour will cause a system to fail in a business sense, the behavioural description must be included. For example, if a securities house bases its buy/sell decisions on the calculation of a share index, the algorithm for calculation of the index must be provided in detail.

- **Designability** - the designer should be able to take the statement of requirement and produce a design with minimal contact with the customer.

- **Preferred modes of operation** - the user may expect the system to behave in a certain way. In this case the description of the user interface shall fall within the scope of requirements definition. For example, an air traffic controller is vitally interested in the modes used by the system to present aircraft coordinates, speed and direction. This will extend to the definition, in the SRS, of exact screen formats, modes of operation, refresh frequencies, data accuracy, data display precision and key stroke sequences. At the other extreme, in the definition of an accounts payable system, the accountant will probably not be concerned with the order in which the cursor moves from one screen data window to another.

- **External system constraints** - if the target system is to become a component of a larger existing system its external behaviour may be heavily constrained by that system. Details of interface with the parent system shall therefore be included in the scope of the SRS. For example, the requirements of a telephone hand set will be constrained by the user's expectation of telephone operation and the standard mode of interface with the telephone network. In this instance a user interface description and a telephone network interface protocol description would be a minimum requirement.

- **Business/legal constraints** - if system behaviour is constrained by externally imposed business or legal rules, the procedures and work instructions shall be referenced or described in detail in the SRS. For example, in the context of a telephone exchange operation, a telephone company policy may state that "no employee shall have the ability to manually adjust a customer's telephone meter". With this in mind a Customer Exchange Requirements Specification would state.

- The system shall increment the calling party's meter when the called party hand set is taken off hook.

- The meter value shall not be modified by any system function other than that described in Requirement 1.

- **Equitable tender evaluation** - if the SRS is to form the basis of a request for tender, sufficient detail must be provided to allow for a fair and reasonable comparison of the cost/performance of each response. If the scope of the tender is design/ implement/ test/ install, all major functional descriptions, data, data flows and capacity and performance requirements must be provided.

- **Touch it, see it, smell it** - in the context of requirements definition, the computer literate requirements engineer typically progresses from real world abstract concepts

45

to the description of physical objects to the definition of computer related logical concepts. Systems engineers tend to drift toward their field of expertise allowing preconceived design approaches to creep into the SRS. As a general guideline, requirements stop the behavioural description of physical objects that we can touch, see and smell. For example we can touch a telephone hand set, we can hear a dial tone and we can see a calling line identification number. We have no physical senses to detect an index, a message terminator or a data structure.

D.3. Content

This section outlines the minimum content of the SRS.

The SRS contains three broad sections. Sections 1 & 2 provide introductory and background information to the target system. Section 3 provides all the details that the software developer needs in order to create the design and is the largest and most important part of the SRS. Section 3 describes the system in terms of the behaviour of its component parts.

While all the generic types of requirements described in the following sections must be addressed, the structure of the document will be determined by whether a specific type of requirement applies to an individual component or to the system as a whole. This makes possible a variety of organisational schemes.

Issues relevant to particular systems that are not provided by this outline may be included as required by the Analyst and approved by the Project Manager.

If a section is not relevant to the target system it must be included and marked 'not applicable' in the body of the text.

In general terms, the behaviour of each component requirement must be described in the following terms:

- Introduction.

- Input.

- Processing.

- Output.

Either individual components or the complete system must also be described as follows:

- External interface requirements.

- Performance requirements.

- Design constraints.

- Security requirements.

- Maintainability requirements.

- Reliability requirements.

- Availability requirements.

- Database requirements.

- Documentation requirements.

- Safety requirements.

- Operational requirements.

- Site adaptation.

The sections which follow indicate the content of each component of the SRS.

D.3.1. Table of contents

The following table indicates the document outline of the SRS. The arrangement of section 3 can vary according to the nature of the software being specified. Examples of variations are given in section 2.4.5 Organising the SRS.

Table of Contents

1. Introduction

    1.1 Purpose

    1.2 Scope

    1.3 Definitions, acronyms & abbreviations

    1.4 References

    1.5 Document overview

2. General description

    2.1 Product perspective

    2.2 Product functions

    2.3 User characteristics

    2.4 General constraints

    2.5 Assumptions & dependencies

3. Specific requirements

    3.1 Functional requirements

Part D: Software Requirements Specification (SRS)

D.3.2. Introduction (section 1)

The purpose of this section of the SRS is to provide an overview of the purpose and scope of the system.

D.3.2.1. Purpose (section 1.1)

Include the following:

- Identify the system name.

- Describe the purpose of the SRS.

- Identify the intended audience for the document.

- Identify the customer and the project team.

D.3.2.2. Scope (section 1.2)

Include the following:

- List the software products described by the SRS.

- Describe what the products will and if necessary won't do.

- Describe the application of the software being specified, including benefits, objectives and goals.

D.3.2.3. Definitions, acronyms and abbreviations (section 1.3)

Include the following:

- Describe terms used that cannot be expected to be common knowledge within the document's target audience.

- Exclude data entity and attribute names that are to be included in data dictionaries.

D.3.2.4. References (section 1.4)

Include the following:

- Identify the preceding documents from which this document was created.

- Identify all documents referenced in this document specifying reference code (to be used in the body of the text), document name, author, version number, release date and location.

- Identify the sources from which the references may be obtained (if applicable).

D.3.2.5. Document overview (section 1.5)

Provide an overview of the remainder of the document describing, in brief, the purpose and content of each level one paragraph.

For example.
Section 2 provides a general description of user requirements.
Section 3 provides specific details of user requirements.

D.3.3. General description (section 2)

This section of the SRS describes the general factors that affect the product and its requirements. It does not state specific requirements, it only makes those requirements easier to understand.

D.3.3.1. Product perspective (section 2.1)

Include the following points.

- Describe (in overview) the business process that the system will support.

- State if the product is self contained (stand alone) or a component of a larger system.

- If the product is a component of a larger system describe its principal functions.

- Identify principal external interfaces.

- Provide an overview of computer hardware to be used if already determined.

- Provide a block diagram describing major components and interfaces.

This subsection should not prescribe specific design solutions or design constraints on the solution. Rather it needs to explain why certain design constraints are specified in section 3.

D.3.3.2. Product functions (section 2.2)

This subsection provides a summary of the functions that the software will perform. It will discuss major functional areas without mentioning any of the large amounts of detail that goes with those areas. The functions need to be organised so that the customer will understand them.

Include the following points:

- A summary of functions to be performed.

- A functional model (i.e. data flow diagram) if appropriate.

- Reasons why specific requirements are stated in Section 3 (i.e. relate the functions to the business processes they support).

Where applicable, the function summary for this subsection can be taken directly from the higher-level specification.

This subsection should not prescribe specific design solutions or design constraints on the solution. Rather it needs to explain why certain design constraints are specified in section 3.

D.3.3.3. User characteristics (section 2.3)

This subsection describes the general characteristics of the users that will influence the SRS.

Describe intended system users in terms of the following:

- Educational level.

- Experience.

- Technical expertise.

Describe the following:

- The operational environment (i.e. physical location, temperature, humidity, vibration, electrical noise).

- Frequency of use (i.e. occasional use, constant use). Provide reasons why specific user-related requirements or design constraints are stated in section 3.

D.3.3.4. General constraints (section 2.4)

This subsection gives a general description of any other items that will limit the developer's design options.

Define limiting factors in the creation of a design.

- Regulatory policies.

- Hardware/system software resource limitations (i.e. memory, disk space).

- External interfaces.

- Audit requirements.

- Company financial control policy.

- Communications protocols.

- Application criticality.

- Safety, security.

- Operations (i.e. unattended operation).

Provide general descriptions only. Further detail specific to the requirements impacted shall be provided in section 3.

D.3.3.5. Assumptions & dependencies (section 2.5)

Define assumed factors that would cause the SRS to change should the assumption be incorrect.

The successful operation of the system is dependant upon the following assumptions and dependencies. Acceptance of this specification means acceptance of the risks associated with these issues.

For example:

- Hardware availability.

- Behaviour of interfacing external systems

- Availability of experienced operations staff.

These assumptions and dependencies are verified individually with the customer. This verification is important in that it helps to arrive at a clear understanding between the developer and the customer which will, in turn, avoid potential contract disputes later.

D.3.4. Specific requirements (section 3)

This section of the SRS provides comprehensive detail on all requirements. It should include all of the detail which the designer will need to create the design. The details shown in section 3 should be defined as individual specific requirements, following the guidelines given in section 2.2 of this standard (i.e. correct, unambiguous, complete).

Background information should be given by cross-referencing each specific requirement to any related discussion in the introduction, general description or appendixes of the SRS.

Requirements are either behavioural or nonbehavioural.

- **Behavioural** requirements describe the functions of the system (functional requirements). They describe the inputs and outputs to and from the system and the processing required to turn inputs into outputs. The bulk of this section is concerned with behavioural requirements.

- **Nonbehavioural** requirements define the attributes that the system exhibits as it performs its job. For example, they include descriptions of efficiency, reliability, security, maintainability, portability, capacity, dynamic performance and standards compliance.

Note: While each of the headings shown in section 3 of the table of contents are mandatory, their organisation can vary according to the nature of the software being specified.

D.3.4.1. Functional requirement n (section 3.1.n)

This section of the standard provides the format of the functional requirement description.

Introduction (section 3.1.n.1)

Include the following:

- The purpose of the function.

- Introductory background material that might clarify the intent of the function.

- Techniques or strategies employed.

- A reference to the source of information that was used as a basis for definition of this requirement. This may be a document name and paragraph number or a reference to a discussion with a user.

- The relative necessity of the requirement. This element provides for negotiation between the customer and the development organisation, preventing situations where the developer spends an inordinate amount of time trying to satisfy a particular requirement when, in the customer's view it could be traded for earlier delivery of the product.

Inputs (section 3.1.n.2)

Include the following:

- A detailed description of all the data input to this function as follows.

  - the sources of the inputs
  - quantities
  - units of measure
  - timing (i.e. the rate of arrival of the data at the system boundary)
  - range of valid input values
  - data accuracies/tolerances
  - the physical format of the data (i.e. paper form, floppy disk, document).

- Details of operator control requirements and activities (i.e. in the context of cheque printing - operator verification that the cheque number loaded on the printer is the cheque number expected by the system).

- Cross-references to interface specifications or interface control documents where appropriate.

Processing (section 3.1.n.3)

Provide descriptions of all operations to be performed on input data and intermediate parameters to obtain the output.

Include the following:

- Validity checks on input data.

- The exact sequence of operations to include the description and timing of events that trigger processing (i.e. purchase requisition raised by maintenance foreman - purchase order created by purchasing department - .....). If synchronism with other processes is important it should be stated here.

- Responses to abnormal situations such as.
  - overflow
  - communication failure
  - power failure
  - invalid data input

- Parameters affected by the operations.

- Requirements for degraded operation (i.e. functions to be provided by an automatic teller machine when the central computer has failed).

- Processing descriptions, for example.
  - equations
  - mathematical algorithms (i.e. data encryption)
  - logical operations
  - modelling techniques (i.e. a process control engineer might require that
  - a missile guidance system be modelled as a finite state machine)

- Validity checks on output data

Outputs (section 3.1.n.4)

Include the following:

- A detailed description of all data output from this function including.
  - destinations of the outputs
  - quantities of data
  - units of measure
  - timing
  - range of valid outputs
  - accuracies/tolerances

- disposition of illegal values
- error messages.

▪ Cross-references to interface specifications or interface control documents where appropriate.

D.3.4.2. External interface requirements (section 3.2)

This section of the SRS describes the modes in which people and external hardware and software will interact with the system.

It is recommended but optional to produce an Interface Control Document (ICD) to specify the requirements for the following interfaces.

▪ Hardware

▪ Software

▪ Communications

▪ User interfaces (section 3.2.1)

▪ Include the following:

▪ The characteristics that the software must support for each human interface to the product. Examples of user interfaces include.
  - menu screen formats and hierarchies
  - data entry screen formats
  - page layout and content of reports
  - relative timing of inputs and outputs
  - function key descriptions

- interface device requirements (i.e. light pens, laser bar code scanners, mice).

▪ Descriptions of the optimisation of interfaces for manipulation by all system users (i.e. command driven for frequent users and menu driven with help messages for infrequent users).

▪ Quantitative measures of user interface quality (i.e. "the Purchase Order Entry System shall be operable by a Clerk Grade 3 with one week's training").

Hardware Interfaces (section 3.2.2)

Include the following:

▪ Specifications of the logical characteristics of each interface between the software product and the hardware components of the system. For example.
- devices to be supported (i.e. printers, terminals, data acquisition equipment)
- technical details of hardware devices (i.e. register addresses and formats,
   protocols).

Software Interfaces (section 3.2.3)

Include the following:

▪ Identification of existing or purchased software to be interfaced with this product. For example: i. Operating system ii.Data management system iii.Mathematical package

Note: this information is provided here if and only if the implementation of the specified software products is constrained by the customer (i.e. not to be left to the discretion of the designer).

- For each software product specify.
    - name
    - mnemonic
    - specification number
    - version number
    - source

- For each interface specify.
    - purpose
    - protocol
    - message format and content.

Note: if the interface is well documented elsewhere provide a reference and/or include the interface description in an appendix.

Communications interfaces (section 3.2.4)

If the system must implement external data communications protocols details or references shall be provided here.

D.3.4.3. *Performance requirements (section 3.3)*

This section shall quantify both static and dynamic performance requirements placed on the software and on human interaction with the software.

Include the following:

- Static requirements in terms of.
  - number of physical input devices to be supported (i.e. terminals, analog/digital data acquisition channels, bar code scanners)
  - number of concurrent users to be supported
  - volume of data to be stored
  - maximum size of data stores (i.e. files, tables)
  - the number of data stores
  - spare data storage capacity required, if any.

- Dynamic requirements in terms of.
  - transactions per second (by type)
  - data acquisition rate (i.e. 4000 analog values per second)
  - the volume of data to be processed in specified time intervals
  - the maximum permissible delay in response to specified stimuli (i.e. "the lift control system shall be capable of tripping the emergency breaking system within two (2) seconds of the receipt of a lift over-speed alarm input")

D.3.4.4. Design constraints (section 3.4)

This section of the SRS specifies all factors that shall constrain the design. They are usually non-negotiable items that shall not be left to the imagination of the designer.

If applicable specify.

- References to design standards such as.
    - design methodology and notation
    - report format
    - data naming conventions.

- Company policy impacting the system. For example.
    - accounting procedures
    - audit requirements (i.e. audit trails)
    - standard procedures and work instructions.

- Hardware limitations imposed on the designer (i.e. "the system shall
    - 1) be implemented on a Motorola 6809 processor and
    - 2) be implemented in a maximum of 64 kilobytes of nonpaged memory").

- Software limitations imposed on the designer (i.e. "in order to preserve consistency with existing applications the system shall be implemented with the XYZ relational database system").

D.3.4.5. Security requirements (section 3.5)

Specify requirements for the protection of software and data from accidental or malicious access, use, modification, destruction or disclosure. Required strategies may include the following:

- Specification of cryptographical techniques

- Provision for logging of system access

- Password security by user and/or function

- Restriction of communication between system functions

- Physical security (i.e. locked doors)

- Computation of checksums for critical quantities.

D.3.4.6. Maintainability requirements (section 3.6)

Maintainability requirements concern the organisation that will maintain the system throughout its life. Maintainability requirements may be expressed in terms of the following:

- Specific coupling metrics for software modules, for example.
  - the maximum number of data items that may be passed between functions
  - the period of time that a component of a system can operate without interface with other components.

- Testability - the ability to regression test system functions in isolation. This may extend to the statement of requirements for regression test strategies, test harnesses and test beds.

- Mean time to repair - the expected time required to repair a system and return it to normal operations.

- Error instrumentation requirements - requirements of software to be embedded in the application to aid in the discovery and tracing of faults and performance problems.

D.3.4.7. Reliability requirements (section 3.7)

If applicable provide quantification of reliability in terms of the following:

- Mean time between failures (MTBF) - the minimum tolerable time between consecutive failures of the system.

D.3.4.8. Availability requirements (section 3.8)

Provide the required system availability in terms of metrics such as the following:

- Hours per day.

- Specific time periods during the day.

- Percent availability averaged over days/weeks/months/years.

D.3.4.9. Database requirements (section 3.9)

Specify the requirements of any database that is to be developed as part of the product.

Include the following:

- Data element descriptions.

- Relationships between data elements.

- Frequency of use.

- Common data accesses (i.e. common database queries).

- Data retention requirements.

- If an existing database package is to be implemented, instructions for its use shall be stated or referenced here.

D.3.4.10. Documentation requirements (section 3.10)

Specify the system documentation that is to be delivered with the product to aid the customer in product installation, user training, product maintenance and system operation.

D.3.4.11. Safety requirements (section 3.11)

Special software requirements originating from safety considerations shall be grouped here. Safety considerations relate to the potential for injury of a human being and/or the damaging or destruction of data and equipment. Requirements could include the following:

- Failsafe operation - the requirement of a system to fail in a predictable, safe way (i.e. in the context of a process control system; "on computer halt all digital outputs shall de-energise all final control elements").

- Interlocking of operator commands.

- Limitation of access to potentially damaging commands (i.e. in the context of a process control system; the setting of pressure, temperature and flow trip limits on pumps, compressors and heat exchangers).

- The ability to manually override computer outputs.

D.3.4.12. Operational requirements (section 3.12)

This section specifies the normal operations required by the user. Depending on the extent of the operations these requirements may be stated here or incorporated into section 3.2.1 (User Interfaces). Operational requirements may include the following:

- Modes of operation (i.e. batch, interactive, system initiated, user initiated).

- Periods of interactive operation and periods of unattended operations.

- Required data processing support functions.

- Backup and recovery operations.

D.3.4.13. Site adaptation (section 3.13)

If applicable provide the following:

- The requirements for any data or initialisation sequences that are specific to a given site, mission or operational mode (i.e. safety limits, grid values).

- A specification of the site or mission-related features that should be modified to adapt the software to a particular installation.

D.3.5. Organising the specific requirements

Section 3 is usually the largest and most complex part of the SRS. its complexity makes it necessary to subdivide this section to most clearly reflect the primary classes of functions to be performed by the software.

The purpose of subdividing this SRS section is to improve clarity and readability, not to define the high-level design of the software being specified. Hence there is no one prescribed way to subdivide section 3. The decision will be determined by the nature of the software being specified. Provided no headings are left out, the arrangement is flexible. The following four examples show possible subdivision variations.

Example 1 - All the functional requirements are specified, then the four types of interface requirements are specified, followed by the remaining requirements. This example reflects the standard pattern shown in the table of contents section:

3 Specific Requirements
    3.1 Functional Requirements
        3.1.1 Functional Requirement 1
            3.1.1.1 Introduction

        3.1.1.2 Inputs
        3.1.1.3 Processing
        3.1.1.4 Outputs
      3.1.2 Functional Requirement 2

      . . .

   3.2 External Interface Requirements
      3.2.1 User Interfaces

      . . .

   3.3 Performance Requirements

Example 2 - Each functional requirement has the complete list of subordinate aspects arranged beneath them, treating each requirement in a self-contained way.

In practice, this is a very workable arrangement and is recommended.

3 Specific Requirements
   3.1 Functional Requirements
      3.1.1 Functional Requirement 1
        3.1.1.1 Introduction
        3.1.1.2 Inputs
        3.1.1.3 Processing
        3.1.1.4 Outputs
        3.1.1.5 External Interface Requirements
          3.1.1.5.1 User Interfaces
          3.1.1.5.2 Hardware Interfaces
          3.1.1.5.3 Software Interfaces
          3.1.1.5.4 Comms Interfaces
        3.1.1.6 Performance Requirements
        3.1.1.7 Design Constraints

. . .

3.1.2 Functional Requirement 2

Example 3 - Section 3 is partitioned into the major areas of functionality. This format separates areas of concern so that different readers can easily find the parts of the specification that are relevant to their own areas of interest.

3 Specific Requirements
    3.1 Functional Requirements
        3.1.1 Functional Area 1
            3.1.1.1 Functional Requirement 1
                3.1.1.1.1 Introduction
                3.1.1.1.2 Inputs
                . . .

            3.1.1.2 Functional Requirement 2
      . . .
        3.1.2 Functional Area 2
            3.1.2.1 Functional Requirement 1
            3.1.2.2 Functional Requirement 2
    . . .
    3.2 External Interface Requirements
    3.3 Performance Requirements
    . . .

Example 4 - Shows the four classes of interface requirements applied to each individual functional requirement. This followed by the rest of the requirements.

3 Specific Requirements
    3.1 Functional Requirements
        3.1.1 Functional Requirement 1
            3.1.1.1 Specification
                3.1.1.1.1 Introduction
                3.1.1.1.2 Inputs
                3.1.1.1.3 Processing
                3.1.1.1.4 Outputs
            3.1.1.2 External Interfaces
                3.1.1.2.1 User Interfaces
                3.1.1.2.2 Hardware Interfaces
                3.1.1.2.3 Software Interfaces
                3.1.1.2.4 Comms Interfaces
        3.1.2 Functional Requirement 2
        . . .
    3.2 Performance Requirements
    3.3 Design Constraints

D.4. Organising the specific requirements

Section 3 is usually the largest and most complex part of the SRS. its complexity makes it necessary to subdivide this section to most clearly reflect the primary classes of functions to be performed by the software. There is no one optimal organisation for all system types. What follows gives guidance to help select the optimal organisation for the system under consideration.

D.4.1. Organise by mode: version 1

Systems behave differently depending on their *mode* of operation. For example a control system may have different sets of functions depending on its mode -- training, normal, emergency. When organising for system mode, use the outline given in Section 4.1 or 4.2 of this document, depending on whether interfaces and performance are dependant on mode.

3. Specific requirements
    3.1 External interface requirements
        3.1.1 User interfaces
        3.1.2 Hardware interfaces
        3.1.3 Software interfaces
        3.1.4 Communications interfaces
    3.2 Functional requirements

3.2.1 Mode 1

    3.2.1.1 Functional requirement 1.1

    .

    .

    3.2.1.n Functional requirement 1.n

3.2.2 Mode 2

.

.

3.2.m Mode m

    3.2.m.1 Functional requirement m.1

    .

    .

    3.2.m.n Functional requirement m.n

3.2 Functional requirements

3.3 Design constraints

3.4 Software design attributes

3.5 Other requirements

D.4.2. Organise by mode: version 2

See version 1 for applicability statement.

3. Specific requirements
    3.1 Functional requirements
        3.1.1 Mode 1
            3.1.1.1 External interfaces
                3.1.1.1.1 User interfaces
                3.1.1.1.2 H'ware interfaces
                3.1.1.1.3 S'ware interfaces
                3.1.1.1.4 Communications interfaces
            3.1.1.2 Functional requirements
                3.1.1.2.1 Functional req 1

                .
                .

                3.1.1.2.n Functional req n
            3.1.1.3 Performance requirement
        3.1.m Mode m
        .

        .

3.3 Design constraints
3.4 Software design attributes
3.5 Other requirements

D.4.3. Organise by user class

Systems behave differently depending on the class of user. For example an elevator control operates differently for passengers, maintenance workers and fire-fighters.

3. Specific requirements
    3.1 External interface requirements
        3.1.1 User interfaces
        3.1.2 Hardware interfaces
        3.1.3 Software interfaces
        3.1.4 Communications interfaces
    3.2 Functional requirements
        3.2.1 User class 1
            3.2.1.1 Functional requirement 1.1

           .

           .

            3.2.1.n Functional requirement 1.n
        3.2.2 User class 2

    .

    .

        3.2.m User class m
            3.2.m.1 Functional requirement m.1

           .

           .

            3.2.m.n Functional requirement m.n
    3.2 Functional requirements
    3.3 Design constraints
    3.4 Software design attributes
    3.5 Other requirements

D.4.4. Organise by object

Objects being real-world entities that have a system counterpart. For example in a health informatics system, objects include patients, doctors, nurses, rooms, medication etc. Each object has associated with it a set of attributes and functions performed by the object.

3. Specific requirements
    3.1 External interface requirements
        3.1.1 User interfaces
        3.1.2 Hardware interfaces
        3.1.3 Software interfaces
        3.1.4 Communications interfaces
    3.2 Classes/objects
        3.2.1 Class/object 1
            3.2.1.1 Attributes - direct or inherited
                3.2.1.1.1 Attribute 1

                .

                .

                3.2.1.1.n Attribute n
            3.2.1.2 Functions (services- direct or inherited)
                3.2.1.2.1 Functional requirement 1.1

                .

                .

                3.2.1.2.m Functional requirement 1.m
            3.2.1.3 Messages (comms received or sent)
        3.2.2 Class/object 2

        .

.

3.2.p Class/object p

3.3 Performance requirement

3.4 Design constraints

3.5 Software design attributes

3.6 Other requirements

D.4.5. Organise by feature

A feature is an externally desired service to be performed by the system, often requiring input data with which to produce the required result. For example, a telephone system features local call, call forwarding, call waiting, voice-mail etc. Each feature can be described by a sequence of stimulus-response pairs.

3. Specific requirements
    3.1 External interface requirements
        3.1.1 User interfaces
        3.1.2 Hardware interfaces
        3.1.3 Software interfaces
        3.1.4 Communications interfaces
    3.2 System features
        3.2.1 System feature 1
            3.2.1.1 Introduction/purpose of feature
            3.2.1.2 Stimulus/response sequence
            3.2.1.3 Assoc. functional reqs
                3.2.1.3.1 Functional req 1

                .
                .

                3.2.1.3.n Functional req n

                .
                .

                3.2.1.1.n Functional req n
        3.2.2 System feature 2

        .

        .

3.2.m System feature m

.

.

3.3 Performance requirement

3.4 Design constraints

3.5 Software design attributes

3.6 Other requirements

D.4.6. Organise by stimulus or response

Some systems are best described in terms of stimuli. For example, an automatic aircraft landing system may be organised into loss of power, wind-shear, sudden change in roll, vertical velocity excessive.

Other systems are best described in terms of the functions that support the generation of a response. For example, a human resources system may be organised into sections corresponding to all functions relating to payroll, current employee list etc.

3. Specific requirements
    3.1 External interface requirements
        3.1.1 User interfaces
        3.1.2 Hardware interfaces
        3.1.3 Software interfaces
        3.1.4 Communications interfaces
    3.2 Functional requirements
        3.2.1 Stimulus 1
            3.2.1.1 Functional requirement 1.1
            3.2.1.2 Functional req 1.2
            .

            .

            3.2.1.n Functional req 1.n
        3.2.2 Stimulus 2
        .

        .

        3.2.m Stimulus m

3.3 Performance requirement
3.4 Design constraints
3.5 Software design attributes
3.6 Other requirements

D.4.7. Organise by functional hierarchy

Where none of the above apply, Section 3 of the SRS can be organised into a hierarchy of functions according to common inputs, common outputs, or common internal data access. Data flow diagrams and data dictionaries can be used to show the relationship between the functions and data.

3. Specific requirements
   3.1 External interface requirements
      3.1.1 User interfaces
      3.1.2 Hardware interfaces
      3.1.3 Software interfaces
      3.1.4 Communications interfaces
   3.2 Functional requirements
      3.2.1 Information flows
         3.2.1.1 Data flow diagram 1
            3.2.1.1.1 Data entities
            3.2.1.1.2 Pertinent processes
            3.2.1.1.3 Topology
         3.2.1.2 Data flow diagram 2
            3.2.1.1.1 Data entities
            3.2.1.1.2 Pertinent processes
            3.2.1.1.3 Topology
            .
            .
         3.2.1.n Data flow diagram n
            3.2.1.n.1 Data entities
            3.2.1.n.2 Pertinent processes

3.2.1.n.3 Topology

3.2.2 Process descriptions

3.2.2.1 Process 1

3.2.2.1.1 Input data entities

3.2.2.1.2 Algorithm or formula of process

3.2.2.1.3 Affected data entities

3.2.2.2 Process 2

3.2.2.2.1 Input data entities

3.2.2.2.2 Algorithm or formula of process

3.2.2.2.3 Affected data entities

.

.

3.2.2.m Process m

3.2.2.m.1 Input data entities

3.2.2.m.2 Algorithm or formula of process

3.2.2.m.3 Affected data entities

3.2.3 Data construct specification

3.2.3.1 Construct 1

3.2.3.1.1 Record type

3.2.3.1.2 Constituent fields

3.2.3.2 Construct 1

3.2.3.2.1 Record type

3.2.3.2.2 Constituent fields

.

.

3.2.3.p Construct p

3.2.3.p.1 Record type

3.2.3.p.2 Constituent fields

3.2.4 Data dictionary

3.2.4.1 Data element 1

3.2.4.1.1 Name

3.2.4.1.2 Representation
3.2.4.1.3 Units/format
3.2.4.1.4 Precision/accuracy
3.2.4.1.5 Range
3.2.4.2 Data element 2
3.2.4.2.1 Name
3.2.4.2.2 Representation
3.2.4.2.3 Units/format
3.2.4.2.4 Precision/accuracy
3.2.4.2.5 Range
.
.
.
3.2.4.q Data element q
3.2.4.q.1 Name
3.2.4.q.2 Representation
3.2.4.q.3 Units/format
3.2.4.q.4 Precision/accuracy
3.2.4.q.5 Range
3.3 Performance requirement
3.4 Design constraints
3.5 Software design attributes
3.6 Other requirements

D.4.8. Organise showing multiple organisations

A combination of any of the above can be tailored to meet your specific needs. The guiding principle is that section 3 be logical and clear, such that it can be readily understood by the customer as well as the supplier.

3. Specific requirements
    3.1 External interface requirements
        3.1.1 User interfaces
        3.1.2 Hardware interfaces
        3.1.3 Software interfaces
        3.1.4 Communications interfaces
    3.2 Functional requirements
        3.2.1 User class 1
            3.2.1.1 Feature 1.1
                3.2.1.1.1 Introduction/purpose of feature
                3.2.1.1.2 Stimulus/response sequence
                3.2.1.1.3 Associated functional requirements
            3.2.1.2 Feature 1.2
                3.2.1.2.1 Introduction/purpose of feature
                3.2.1.2.2 Stimulus/response sequence
                3.2.1.2.3 Associated functional requirements
                .

                .

            3.2.1.m Feature 1.m
                3.2.1.m.1 Introduction/purpose of feature
                3.2.1.m.2 Stimulus/response sequence

3.2.1.m.3 Associated functional reqs

3.2.2 User class 2

.

.

3.2.n User class n

.

.

3.3 Performance requirement

3.4 Design constraints

3.5 Software design attributes

3.6 Other requirements

E. References

[1] Scaacchi, W., (2004) *The Encyclopedia of Human-Computer Interaction*, W. S. Bainbridge (ed.) Berkshire Publishing Group, 2004

[2] Grindley, K. (1991). Managing IT at board level: The hidden agenda exposed. London, Pitman

[3] Edstrom, A. (1977). User influence and the success of IS projects: a contingency approach. Human Relations, 30(7), pp.589-607.

[4] Gingras, L., & McLean, E.R. (1979). *A study of users and designers of information systems* (IS working paper 2-79). California: Graduate School of Management, UCLA

[5] Zmud, R.W. & Cox, J.F. (1979). *The implementation process: a change approach*. MIS Quarterly, 3(2), pp.35-43

[6] Wang, C.B. (1994*). Techno Vision*, New York, McGraw-Hill, p1.

[7] Hornik, S., Chen, H-G, Klein, G., Jiang, J., (2003). *Communication Skills of IS Providers*: An Expectation Gap Analysis From Three Stakeholder Perspectives. IEEE Transactions on Professional Communication, Vol 46, No. 1. pp 17-29.

[8] Mann, J., (2002). IT Education's Failure to Deliver Successful Information Systems: Now is the Time to Address the IT-User Gap. Journal of Information Technology Education, Vol 1, No. 4, pp 255-256.